*W*ater is a traveler on the blue planet. Flowing water carves monuments from rock, alters boundaries between land and sea and, frozen in the form of glaciers, scours the landscape. Water weaves together all living and nonliving things in an intricate tapestry of color, light, and movement. So much a part of us and our routines that we often take it for granted and forget that water—a gift of nature—is life.

In abundance or scarcity, water creates diversity on the earth. Water is often referred to as a sculptor, but it is also a painter. With masterful strokes water brings the colors of life to earth.

A rainbow after a summer storm; horses racing overhead in the ever-changing shapes of clouds; the taste of cold water on a hot day—sharing experiences, we come together as a world community in our appreciation of and need for clean and plentiful water.

National Project WET (Water Education for Teachers) is an interdisciplinary water education program that believes the study of water should embrace natural and social sciences, mathematics and the arts. Designed for young people in preschool through high school, teachers, parents, and youth leaders, National Project WET is grounded in the belief that when informed, people are more likely to participate in the decision-making process and to make a difference through their actions.

In support of its beliefs, National Project WET is developing a curriculum guide and water modules; children's water story and action books; the Groundwater Flow Model; the Liquid Treasure Trunk (a water history project); and the Watershed Management Simulator.

Funded by the U.S. Department of Interior, Bureau of Reclamation, National Project WET is looking forward to sponsorship in all 50 states and international involvement. National Project WET invites you to share in a celebration of one of the most precious resources on the planet—water.

Dennis Nelson, Director
National Project WET

Edited by Mary L. Van Camp
Book design by K. C. DenDooven

First Printing, 1993
WATER—A GIFT OF NATURE: THE STORY BEHIND THE SCENERY
© 1993 KC PUBLICATIONS, INC.

LC 93-77028. ISBN 0-88714-077-7.

Front cover: Lower Falls in the Grand Canyon of the Yellowstone, photo by Larry Ulrich. Inside front cover: Cook Lakes—headwater of the Green River, Wind River Range, Wyoming, photo by Gary Ladd. Page 1: Icebergs on a glacial lake in Jokulsarlon, Iceland, photo by C. Allan Morgan. Pages 2/3: Proxy Falls, Oregon, photo by Dick Dietrich. Pages 4/5: Great Fountain Geyser in Yellowstone National Park, photo by Glenn Van Nimwegen.

WATER *A Gift of Nature*
THE STORY BEHIND THE SCENERY®

by Sandra Robinson, Dennis Nelson, Susan Higgins, and Michael Brody

Sandra Robinson, writer/editor for National Project WET, is an award-winning author. Dennis Nelson, Director of National Project WET is a leader in water resources education. Susan Higgins is a former state water planner. Michael Brody, Ph.D., Associate Director, National Project WET, has researched children's understanding of ecology. Special thanks to Linda Hveem and Nancy Carrasco.

Water's Incredible Journey

Watching the sun's final rays wash the sea in a golden glow, a child might ask, "Where does all the water come from?" Early migrating people who pushed up against the "blue boundary" knew that the rivers they followed emptied into the sea, but how did the waters get into the rivers?

The pathway that water follows is called the hydrologic cycle. Powered by the sun, water that is evaporated from oceans, rivers, lakes, and soil or transpired from plants rises into the air. Cooled in the atmosphere, water condenses. Falling as rain, snow, or hail to the earth, water flows over and through the land. In the return to the sea, the waters rush or meander in rivers, their character determined by their course.

In its incredible journey through the cycle, a water drop may assume many roles. Carrying dissolved nutrients from the soil, it may nurture plants. It may quench the thirst of a child or a song bird. It may act as a solvent or a reactant in industrial systems. Therefore, the drop may return to the water cycle with "extra baggage" or through evaporation or treatment may be cleansed.

Civilizations have prospered or faltered in response to water availability. Calibrating the stages of the Nile River an early naturalist related water level and its effect on people. Water depth was gauged not only in "ells" (a form of measurement) but also in terms like "hunger," "happiness," "abundance," and "disaster."

Responding to the demands of our twentieth century lifestyle, our relationship to the water cycle has grown increasingly complex. And yet, it is still as simple as satisfying our need for a clean drink of water.

◁ **Tremendous** energy is released by raindrops. In driving rainstorms, in excess of 100 tons of dirt per acre can be loosened by droplets falling on poorly protected soils. Over time through runoff, these sediments are carried into streams and rivers that find their way to the sea. Beach sands that you sift through your fingers may once have been part of a mountain top.

◁ **As the sun slowly sets it provides a breathtaking** sion of highlighted clouds and golden reflections the Pacific Ocean.

American Falls on the Niagara River is a convincing display for millions of visitors of the energy contained in flowing water. After rain falls on the land it may seep into the ground and become groundwater, be evaporated from the land, drain into rivers and streams, or thunder over ▽ rocky ledges on its path to the sea.

From the Snow

Water covers about 70 percent of the earth's surface. Even so, some communities throughout the world are either experiencing a freshwater shortage or anticipating one.

How can there be scarcity in such seeming abundance? Imagine that the water in your bathtub (about 26 gallons) represents all the water in the world. But 97 percent of the world's water is salt water. Draining your tub of this salt water leaves about one gallon of freshwater, however, only a small percentage of the freshwater is "available" for our use. (A large amount of freshwater is locked up in glaciers, polar ice caps, the atmosphere and the soil, or is too polluted or at depths so great under the earth it is too expensive to remove.) So how much useable water is left in the tub? About one-half teaspoon!

Still there is plenty of freshwater for each person on the planet. Through the water cycle, water is cleansed and distributed. However, with pollution and the increasing demands placed on our usable water we must be more discriminate about how much "goes down the drain."

◁ **Early philosophers believed that snow** occurred when air became locked inside of water. Snow does contain less water than rain. In fact, ten inches of wet snow has about the same amount of water as one inch of rain. Therefore, snowpack—functioning like a "water bank"—is important in the water cycle. When snow in the high country begins to melt in late spring, it is a major source of runoff swelling streams and rivers and recharging groundwater.

◁ **The water cycle connects all living** and nonliving things on the earth. Plants, too, take from and contribute to the hydrologic cycle. Of herbaceous (green and leafy) plants water makes up 80 to 90 percent of their weight—woody plants, about 50 percent. Where temperature favors plant growth, the availability of water is one of the main factors that determines the distribution of plants. Humans perspire and plants "transpire." Through transpiration, plants give off water through pores on their leaves. Plants give up 95 percent of the water they absorb, the remaining 5 percent is used for growth and maintenance. The moist environment of this stream in Alaska's Denali National Park supports a lush growth of vegetation.

Wetlands are ▷ transition areas between land and water. Wetlands range from prairie potholes on the North Dakota plains to the river of grass in south Florida. The saturated soils of wetlands support plant life that can tolerate low levels of oxygen. The tubers of cattails are a preferred food of the common freshwater wetlands inhabitant, the muskrat.

△ **On their way to the sea, streams and rivers, conforming to diverse topography, may plunge over ledges to** create spectacular waterfalls. Where there are bands of soft and hard rock, the force of a fast-moving river wears away the soft rock and leaves a shelf of hard rock over which the water falls. The moist environment is conducive to the growth of dense mats of rich, green moss.

△ **Viewing the quiet splendor of Peyto Lake in Banff National Park, Canada, you witness the impact of water as** *glacial ice. The lake is a remnant of a glacier which, as it melted and retreated uphill, deposited a mound of rock debris. As the glacier continued to melt, its waters dammed up behind the mound and created this beautiful alpine lake. The debris, or terminal moraine, is now a tree-covered hill.*

Water Color

Not only the state—solid, liquid, or gas—but also the appearance of water varies. Water in a lake, pond, or ocean may look green, blue, even brown. The color of water is determined by the interaction of light and water molecules, bottom sediments, dissolved matter, suspended solids, and algal blooms. Even time of day has an effect as the angle of sunlight striking the water varies.

The vivid blue waters of Crater Lake were ▷ *thought to have healing powers by the Klamath Indians. Crater Lake National Park in southwestern Oregon was established in 1902 for the protection of this natural wonder.*

Old Faithful Geyser? ▷
Yes! This small spouter in Napa Valley, California, shares the name of the famous geyser in Yellowstone National Park, but not the attention of 3 million visitors a year. A geyser has three main components: an ample water supply, an underground heat source, and a subterranean reservoir and network of cracks and channels. Rain and snow provide the first requirement. Following underground channels that reach far into the earth, seeping water finally reaches "hot rocks." The heated water rises and flows into the geyser's "plumbing system." This interconnected system, that is unique to each geyser, shapes the appearance of the eruption and the geyser's overall behavior.

Water Magic

In the physical world, water is the "Great Houdini." When it turns to vapor, water seems to "vanish into thin air." With their intricate and individual patterns, snowflakes have a magical quality. However, the behavior of water and its ever-changing appearance have nothing to do with magic.

The chemical structure of water—two atoms of hydrogen and one atom of oxygen—and their relationship determines the character of water. Water exists in three states: solid, liquid, or gas. Within the range of the earth's surface temperatures, no other substance on the planet can exist in these three forms.

Whether skating on a frozen pond, experiencing an eruption of Old Faithful Geyser, or watching waves break on shore there is a sense of magic in the personal experiences we have with water.

△ *Water in clouds can take on any number of shapes. A lion? A dragon? What do you see in this cloud?*

◁ *Redwoods thrive in the foggy conditions of California's Pacific coast. Fog can be thought of as a cloud that rests on the earth. When warm, moist air is cooled, the moisture condenses and forms small water drops. Sunlight breaking through the mist gives this scene an otherworldly quality.*

Water: Agent of Change

"The water advanced like a tremendous wall.... Stores, houses, trees, everything was going down in front of it, and the closer it came, the bigger it seemed to grow...." The Johnstown Flood occurred May 31, 1889. Over 2,000 people were killed and more than $10 million worth of property was destroyed in this Pennsylvania town.

Partnered with gravity, water can exert tremendous force. Changes crafted by water are as subtle as a blossom revived by a gentle rain or as catastrophic as a flood that kills thousands. Even the absence of water can effect change. Without water the earth dries and cracks, crops wither and, experiencing thirst, humans and wildlife are forced to adapt, migrate, or perish.

What qualities give water this power to engineer change? The molecular structure of water makes it an almost universal solvent. In our bodies water in blood or lymph carries nourishment to cells and removes waste. Water plays the role of nurturer and cleanser for people, plants, cities—even the planet.

Because of its ability to absorb and hold heat, water in oceans and lakes maintains the thermal stability of the earth. In plant and animal cells water acts as an insulator by protecting heat-sensitive proteins.

Water takes away, but also leaves behind; it depletes but also restores. Water changes form and yet its molecular structure remains the same. Amazing that this particular arrangement of two hydrogen and one oxygen atoms endows water with the qualities that allow it to accomplish such diverse acts as sculpting mountains and sheltering the unborn in the womb.

△ **The features of Grand Canyon National Park are a** testament to the power of flowing water. The Colorado River provides the main force in the downward erosion in the canyon. However, torrential summer rains and heavy winter snowpack have also helped to sculpt this wonder in stone.

△ **The character of water is multiform.** Water freezing in crevices may crack massive boulders. Flowing water has the force to erode and shape landscapes. Water vapor that cools can appear as gently rounded puffs or streak the sky with wisps like the vane of a feather. Clouds can rise to 64,000 feet and produce rain, snow, hail, or thunderstorms.

We generally think of the erosional capacity of ▷ water in "grand" terms, but even in Grand Canyon National Park water creates forms that are intriguing in their subtlety. Although these "pebble towers" give the illusion of height, they are only two inches tall!

△ **Water is a study in contrast. The presence of water (as either a solid, liquid, or gas) or the absence of water** can effect a variety of changes. The beauty and seeming fragileness of this winter scene is deceptive. Freezing water exerts tremendous pressure. In colder climates the temperature may move between freezing and thawing many times during the year. Water that finds its way into the gaps and fissures of rocks may freeze and melt as many as 70 times annually. Freezing water exerts a force of about 30,000 pounds per square inch. This pressure is enough to fracture rock. Trees will also develop cracks from moisture freezing within cells. Rainfall in Death Valley averages about 1.5 inches a year. In the summer, temperatures can climb to 125 degrees F. In the absence of water, the surface dries and cracks—as if it has been stretched too tightly over the landscape.

◁ **Floodwaters** demonstrate tremendous power. The toil of a human lifetime can be swept away in minutes. For example, in 1972 torrential rains from a tropical storm caused rivers to overflow their banks in New York and Pennsylvania. The resulting damage cost 3 billion dollars and left some 15,000 people homeless.

Imagine a hailstone the size of a soccer ball! In 1970 a △▷ hailstone with a record diameter of 7.5 inches fell from the sky. Generally these balls of ice resemble peas or golf balls. Hail can cause extensive damage to people, cars, and crops. Scientists have experimented with cloud seeding and other techniques in order to break up hail storms and prevent the devastation of crops.

Patterns

Water in its diverse forms creates intriguing patterns in nature. Snow crystals exhibit a great variety of shapes and patterns. Yet, every snow crystal has six sides. If you have peeled an onion, then you are familiar with the internal pattern of hailstones which are made of layers of ice. Clouds represent the collection of water droplets or ice crystals. Cloud patterns are constantly shifting and beyond providing us with the childhood game of "guess the shape," they function in weather predicting. From an aerial perspective, tributaries emptying into rivers on their way to the sea form interesting patterns on the landscape. Perhaps the most beautiful pattern associated with water comes from mixing raindrops and sunlight to create a rainbow—consistent throughout the world with its graceful arch, sequence of colors, and message of hope.

***T**he San Juan* ▷
River flows through a maze of tight switchbacks called "goosenecks." Carrying some 30 million tons of silt, sand, mud, and gravel each year, this river flows into Lake Powell, Arizona.

Over great spans of time Navajo Sandstone was formed by the interaction of two agents of change—wind and water. The Navajo is composed of desert sands deposited by wind and sediments from a shallow inland sea that once covered the region. The Navajo is porous, fine-grained and easily crumbled. Sandstone is not safe for climbing, but ▽ because of its porosity, allows for the storage and movement of water.

Erosion

In a sense erosion is a reshuffling of the deck—soil and rock are broken away from one place on the surface of the earth and moved to another. In several ways water acts as an erosive agent. Water that has seeped into the crevices of rocks freezes and, as it expands, exerts tremendous force. Rocks are split apart by this action. Freed, they can be moved to another location. Water is a "mover." Glaciers slowly grinding over the landscape pick up dirt and rocks in their path and carry them along. As the glaciers melt and retreat they leave these materials behind.

Raindrops striking the earth loosen the dirt and particularly on slopes transport it. Water currents move sediments down riverbeds or to the ocean. Waves pounding the coast wear away rock and change shorelines.

Erosion has both positive and negative effects. Erosion can remove productive soil from agricultural lands, but it has also created many natural wonders.

△ **Double Arch is easily accessed by visitors to** Arches National Park in Utah. It is one of 200 arches in the park and is an extraordinary example of artful sculpting by erosive agents.

MARIE MENZIETTI

The energy in waves is mainly generated by wind. Continually shaping the coastline, waves over time reduce ▽ great boulders to sand.

"Water Address"

Water is one of the more common substances on earth. However, it is not distributed equally across the planet. More than any other one environmental factor, the quantity of water determines the amount and type of vegetation in an area. Therefore, this "inequality" contributes to the diversity of plant communities and the animals associated with them.

Because water—or the lack of it—helps to shape the community in which an animal lives, a species' "water address" might be a rainforest, desert, freshwater pond, or tidal pool.

Some animals have amazing relationships with water. Desert nesting sandgrouse fly great distances to water sources, wet their specialized belly feathers, and return with water for chicks in the nest. When a grizzly bear hibernates, it does not take in any water for as long as seven months.

Some animals manipulate their environment in order to satisfy their water and housing needs. Beavers "engineer" dams and alligators dig and maintain water holes.

Most early human settlements depended on the presence of water for drinking, agriculture, defensibility, transportation, and communication. When water sources dried up or were depleted, people were forced to adapt or to move.

There is roughly the same amount of water on the planet today that there has been from the beginning of time. Yet, our demand for useable water steadily increases. Of the planets we have explored, not one has been discovered that can support life as we know it. Our "water address" is the blue planet, and so far a "forwarding address" is not an option.

▲ **Beaver dams are constructed of a network of** branches cemented with mud that the beaver removes from the stream bottom. The beaver's engineering achievements reduce erosion; hold water in arid parts of the west; create habitat for other species; and increase "browse" for moose, elk, and deer. With nose and ear openings that close under water, an extra eyelid that allows underwater vision, and webbing on its hind feet, the beaver is adapted for an aquatic life.

Like the beaver, the alligator too "engineers" a home for wildlife. Prior to people's manipulation the plants and animals of the everglades were sensitively tuned and adapted to the cycle of wet and dry. During the rainy season wildlife was dispersed throughout the everglades, but as early winter approached the rains stopped. Through evaporation and runoff, the shallow river began to dry. Soon water could only be found in deep sloughs or in depressions maintained by an unlikely caretaker—the alligator.

Florida red-bellied turtle.

Green-backed heron.

As the glades dried in the winter season, life focused in the alligator hole. As fish and frogs congregated in these watery havens, other animals such as green-backed herons or Florida red-bellied turtles followed for food and water. This concentration of food sources coincided with the nesting season of wading birds. Today scientists are trying to re-create this delicate balance. We can only hope to imitate what has been the experience of alligators for a long, long time.

Overleaf: Sunlight revealing the color of raindrops, this double rainbow arches over Grand Canyon National Park. Photo by Tom Till.

PHOTOS BY PETER HOWORTH

△ **Achieving spectacular leaps, humpback whales live in all the world's oceans. Cetaceans (animals such as whales and porpoises) are exceptional for being the mammals best adapted to aquatic life.** Humpback whales are baleen feeders. They have no teeth, but instead have plates of baleen (which looks like combs with closely spaced teeth) through which water passes with krill and small fish filtered out. Krill are tiny shrimplike animals ranging in size from three-eighths of an inch to six inches. Remarkable that such small creatures could support a massive animal like the whale.

◁ **Most scientists believe** that life began in the sea. Ocean water contains all elements essential for life. Regardless of the numerous theories regarding the origin of life, water played a major role in our beginnings. Cells are composed of 60 to 70 percent water. Living things require water for nourishment and the removal of wastes. Sea stars live in this life-giving bath all over the world.

The Wet Side

▲ **M**ainly a fish eater, the great egret depends upon seasonal waters for food and protection. These birds often nest on islands in bays where the expanse of water separates them from most predators.

Animals must strive to ▲ survive in harsh environments, but conditions that test their strength in winter may ensure their survival in spring. Elk that endure winter will graze in spring meadows of rich, green grass fed by snowmelt and gentle rains.

Most species of wildlife are unable to manage their water supply. Over time, however, many organisms have adopted behaviors that respond to the lack or abundance of water. Elk, moose, and deer migrate, sometimes great distances, from summer to winter range in order to avoid heavy snows in the mountains. In level terrain, deer and elk restrict their movements to smaller areas (called "yards"). Moving on established trails, instead of wading through deep snow, they conserve energy.

In south Florida the nesting patterns of birds correlate with the winter dry season. In order to satisfy the food demands of nestlings, parent birds feed on fish congregated in shrinking water holes during the dry season. However, human settlement and development have altered historical water patterns, and wading bird populations have severely suffered.

▲ **The desert landscape is constantly** changing. The flat-topped hills that are generally in the scenery of old western movies are called "mesas." Smaller mounds are named "buttes." The piles of sand arranged by wind are "dunes." The image of the flower and the dune symbolizes the tenacity of life in the desert. Living in the dry West, an early homesteader wrote: "I never said I was hungry unless there was food to eat, and I never said I was thirsty unless there was water to drink." The woman and the flower represent the beauty not of surviving but of blooming in spite of adversity.

The Dry Side

Deserts are stressful environments in which more water may be lost in a year through evaporation and transpiration than is gained through rainfall. Of all places on the planet desert rainfall is the most difficult to predict and the most irregular. Because of adverse conditions such as low rainfall or high temperatures, people once believed deserts were simple systems with little diversity. However, current research maintains deserts are complex and "biologically rich places."

In relation to water, deserts are places of extremes. A desert may experience little precipitation for several years; however, when the rain does come as much as ten inches may fall within a few hours. Desert plants and animals have adapted to this "feast or famine" lifestyle.

△ **The kangaroo rat feeds only on dry seeds.** The rat metabolizes proteins and fats in a manner that provides all the fresh water it requires. With its powerful hind legs the Kangaroo rat can make spectacular leaps up to nine feet.

△ **Imagine a plant able to hold tons of water at one time!** The ribs of the saguaro or giant cactus expand or contract like an accordion in order to store water when it is available. Cacti lose less water through transpiration than other plants because of a thick protective outer covering.

△ **Strictly herbivorous, this desert tortoise feeds** on the bloom of a beantail cactus. Is a tortoise a turtle? Yes, but the tortoise lives only on land.

◁ **Related to the iguana, the** chuckwalla inhabits desert areas of the southwestern United States. Although the chuckwalla is large—growing to one and one-fourth feet—it is harmless to humans. It feeds on tender leaves and flowers and was once used for food by early southwestern people.

The Flowering Desert

This profusion ▷ of flowers shows the response of desert plants to abudant moisture. Few winters at Pichaco Peak State Park, Arizona, are wet enough to produce such a colorful display. After blooming only a few weeks, the plants will scatter their seeds and die. Desert plants cope with water scarcity in many ways. To take advantage of every drop from a brief shower, plants have extensive root systems. Some plants shed their leaves during a drought to avoid loss of moisture through transpiration. Plants, like the cactus, are able to store water in leaves, roots, or stems.

GARY LADD

△ **Throughout history, the availability of water has been a major factor in determining where people live.** The exploration, settlement and development of the United States were largely influenced by water. Rivers provided explorers passage to the wilderness. Cities developed around harbors and rivers where the fruits of commerce could be shipped and delivered. Utilizing the energy of flowing water, mills and factories were built on streams. Today Boston Harbor is a busy seaport.

Drought, an extended period of no rainfall, has forced the migration of families and communities. About A.D. 1000, groups of Pueblo Indians built their houses on protected mountain ledges. Their culture prospered for some 300 years—and then they were gone. Scientists believe their exodus was due to enemy attacks and drought. ▽

Water Works — For All of Us

What do a loaf of bread, a sheet of paper, and an automobile, have in common? WATER! Our use of water is both direct and indirect. In our homes, we bathe, brush our teeth, and flush the toilet. Although it is not obvious to us, large quantities of water grow the grains to make the bread we eat, process wood for paper making, and produce steel used in automobile manufacturing. We are sometimes critical of other water users, however, quite often their "need" for water translated into our "use" of their product.

Water works for us in electric power generation, manufacturing, mining, navigation, fisheries, farming, ranching, and in fulfilling municipal domestic needs. The major rivers of the world support thousands of different water users. The Colorado River is used repeatedly before it flows into the Gulf of California.

Many major water users are looking for ways to reduce their water consumption. Many industries are adopting systems for recycling the water they use. In food production, crops are either watered through irrigation practices or rainfall. Today methods are being developed for more "thrifty" irrigation.

Water is often assessed in economic terms. But how do we measure the "value" of watching a grizzly bear feeding in a mountain stream or a geyser erupting? Experiencing the fragile and exotic beauty of a coral reef or walking the shoreline of a still pond, we acknowledge that beyond economic measures, these waters enrich the human spirit.

△ **Completed in 1942, Grand Coulee Dam on the** Columbia River is the greatest provider of water power in the United States. But dams are not a "new" idea—probably the oldest dam in the world, the Sadd el-Kafara (Dam of the Pagans) was constructed between 2950 B.C. and 2750 B.C.

△ **W**e remove fish and shellfish from the planet's waters to feed both people and livestock. The annual catch is made up of species such as herring, anchovy, tuna, salmon, trout, bass, and perch. Of the 30,000 known fish species, only 40 are harvested in substantial numbers. The shellfish take is mainly lobsters, shrimps, crabs, oysters, clams, mussels, squid, and octopuses. Oceans provide 87 percent of the yearly commercial harvest. Protecting the quality of the oceans is critical if we want to continue to enjoy the "catch of the day."

Water is a function ▷ not only of work but also of "play." Much of our recreation centers around water—water skiing, sailing, swimming, and fishing. Our health depends on the health of our waters.

EARTH SCENES/E.R. DEGGINGER

▲ **A variety of irrigation systems** makes possible the production of crops and fruits in areas that have insufficient rainfall: center-pivot sprinklers (pictured), gravity-flow canals, precision-application sprinklers, and drip irrigation. In drip irrigation water runs through perforated tubes close to crop roots thereby reducing the amount of water lost through evaporation when sprayed into the air.

Water Harvest

After learning to build ▷ primitive boats, people used water for the movement of materials, products, and themselves. Today, heavy, bulky items such as grain, oil, coal, and industrial equipment are moved more economically on water than land.

△ **W**ater "works" in producing an atmosphere for human environments. Fountains have been created for thousands of years. Architects have long realized that people enjoy watching the movement and listening to the sounds of falling water. Designed for libraries, malls, and hotels, some fountains—particularly in water scarce areas—use recycled water.

Everything that we put on our tables is in a sense a "water harvest." Crops would simply not grow in the arid West without irrigation. Irrigation practices that have not been water efficient are being improved by technology. Some farmers use computer-driven systems that identify leaks and release water depending upon soil moisture and weather conditions. Hybrid crops are being developed that are less thirsty or more tolerant of saline waters for irrigation.

The oceans yield a tremendous harvest of fish and shellfish. This feeds people directly and indirectly in that livestock feed is also produced.

Kelp shelters and feeds many marine species, and also feeds people. China and Japan grow kelp for food on special marine farms. The United States gathers it where grown naturally. "Algin" is extracted from kelp and is used in making ice cream, salad dressing, beer, paper, and cosmetics.

Growing only in cold (not tropical) water, "kelps" are brown seaweeds. Some varieties of kelp form forests under the sea. These forests calm the waters and are important habitat for sea otters, lobsters, and many fish. Sea urchins feed on kelp. ▽

△ *In balancing the water equation the needs of all water users must be considered. Without irrigation it would not be possible to raise food in dry places such as the Armagosa Valley, Nevada. The lush, artificially watered farm lands contrast sharply with the surrounding natural landscape.*

Too Little... Too Much

A city of 1 million people in the United States uses about 156,542,617 gallons of water daily. In 1988, almost 75 percent of Americans lived in urban areas. As cities grow, they depend more and more upon outlying sources of water. Expanding water and waste-water systems to meet growing needs is more expensive than instituting a conservation program. In the past, "conservation measures" were taken seriously only during periods of drought. Today by adjusting our attitudes and water-use practices, perhaps "too little" can be "just enough."

During brief intense periods, because of sudden snowmelt, torrential rains, or violent weather conditions, there can be too much water. Flooding can result in loss of life or property. Ironically, floods can be beneficial. The annual floods of the Nile River made the Egyptian Plains one of the most fertile places in the world.

***T**he "All-American" canal carries water to Los Angeles. The water that runs from a tap in Los Angeles has likely traveled hundreds of miles from northern California or the Colorado River Basin. Recognizing the increasing demand on a resource that must be "shared" by many water users, this city has instituted waste-water recycling and other water ▽ conservation practices.*

Although tornadoes are wind events, rain and hail often accompany them. Tornadoes that occur over oceans or bays are called "water spouts." Despite advances in weather manipulation, we cannot manage a tornado—we can only protect ourselves and wait for the storm to pass.

◁ **Although floods are generally** thought of as "natural events," human activities can create environments prone to flooding. Practices that remove soil and vegetation, that act like "sponges" soaking up runoff, can increase an area's vulnerability to flooding. Actions to offset this susceptibility include: replanting vegetation, holding excess water in ponds to be timely released, and diverting excess water to storage tanks for industrial use.

▲ **This bear looks as if it can't believe its good luck.** Hunting migrating salmon, this brown bear fishes the falls in Alaska's Katmai National Park. Katmai is an important bear sanctuary with an estimated 500 brown bears in residence.

▲ **These "fishermen" are both enjoying success.** For the bear it's a matter of survival, but for the angler it's just plain fun! Sport fishing is a multi-billion dollar a year industry, and over 55 million people participate in this sport annually in the United States alone.

◁ **Wind surfing is the union of wind and water to** create an experience that requires a rider's sense of balance and timing. Wind surfers skim the calm waters of Florida Bay off Key Largo, the frigid mountain lakes of the West, and the swells beyond ocean breakers. Without noise or fuel vapors, this sport allows the participant to interact in harmony with the elements he challenges.

△ **O**riginally made of animal skins stretched over a wooden frame, kayaks have been used for hunting and fishing for thousands of years. Kayaking requires sensitivity to the "moods" of wind and water.

△ **W**hether becoming acquainted with brightly colored reef fish, exploring underwater caves, or diving ancient wrecks in search of gold bullion, the enthusiast becomes intimately involved with the resource in the sport of scuba diving.

Recreation — The Need for Good, Clean Water

In the early history of the United States the use of water for recreation was almost nonexistent. An early woman homesteader wrote: "The only water we had for any purpose was contained in a gallon jug, and we did not know how soon nor where we could get more. Consequently, we drank sparingly and in little sips, and bathed our hands and faces in the dewy grass of morning."

Today, in addition to satisfying our physical needs, water quenches our thirst for adventure, diversity, and challenge. Canoeing, kayaking, scuba diving, swimming, fishing, sailing, ice skating, and skiing are popular water sports. But water recreation does not always involve getting "into" water. Bird watchers frequent wetlands, picnickers spread their blankets on lake shores, and sun worshipers flock to coastal beaches.

National parks and preserves are often established for their spectacular "waters." Yellowstone's Old Faithful Geyser, Yosemite Falls, Crater Lake, to name only a few, attract millions of visitors each year.

In poetry water is often characterized as "the wonder of life" or "the gift of life." These expressions are lyrical, but understated, for without water, there is no life!

Our Water Legacy

Regardless of our cultural differences, the need for clean and plentiful water brings us together as a world community. But our demand for water is growing. Since 1950 water use around the world has tripled. Some people believe that we are moving into a period of water scarcity. The social and economic future of our children is not secure unless we can provide clean and abundant water.

What are our options? Certainly, conserving water resources is critical. Many communities are educating citizens about water-saving devices and offering incentives for "learning to live within a water budget." Waste water is being treated to be used for irrigation and city park and lawn watering. Responding to pollution control laws, many industries are recycling water. Irrigation systems are being developed that use water more efficiently.

Researchers are attempting to unlock water resources in clouds, icebergs, and oceans. Some countries have been experimenting with cloud seeding. Thought has been given to moving icebergs to waterless regions to increase freshwater supplies. Desalination plants have been constructed in coastal areas.

In the past we have used the earth's water as if it flowed from the "Miraculous Pitcher," a vessel from which liquids drained without end. We know today that the earth's resources are finite.

Our success or failure as a civilization will depend upon our personal and global commitment to clean and abundant water. Perhaps the best legacy we can leave our children is accepting responsibility for our place in the water cycle.

△ **Stirring a puddle with a stick in a farmyard or a** city lot, listening to the sound of rain on an adobe or shingled roof, delighting in the fragile beauty of dew on a blossom, we share experiences that ignore boundaries and blend cultural differences. To meet the challenge of providing tomorrow's children with clean and plentiful water, we need to respect each other's water requirements and respond as a world community.

△ **For more than 40 years, weather modification efforts, often termed "cloud seeding," have been** deliberately aimed at producing increased rainfall and snowpack, decreasing hail fall, and dissipating fog. Rain and snow are usually increased by seeding suitable clouds with silver iodide or dry ice, which may greatly accelerate the development of rain and/or snow. Hail reductions are also attempted in this manner, as it appears that more efficient storms may also produce less hail, and well-timed, well-placed seeding may increase overall cloud efficiency.

△ **Because the technology exists to remove dissolved salts from seawater, concerns about** freshwater scarcity appear unfounded. However, "desalination" requires great amounts of energy and is expensive. Although desalination is not the solution for all areas, arid coastal cities can benefit where freshwater is at a premium regardless of the method to obtain it.

△ **Aquaculture is the commercial raising and** harvesting of fish, shellfish, and plants. About six percent of the annual fish catch of the world is provided by aquaculture. Tilapia (pictured) is a species raised on fish farms. Maintaining water quality and the integrity of native fish species are challenges faced by aquaculturists.

Living on a Water Budget

Cloud seeding, desalination, towing icebergs, rainwater harvesting, more efficient irrigation, water-saving plumbing devices, and waste-water recycling may all be necessary in order to balance the world water budget. Equilibrium must be maintained between the needs of people and the integrity of ecological systems. How do we approach such formidable issues?

Single raindrops acting over time can erode boulders. A sea full of water droplets can rearrange coastlines. The answers to big questions can have small beginnings—in homes, schools, and offices. The solution begins with the individual. Human attitudes that lead to positive actions can effect major changes.

Water laws dating back to the time of Plato were very progressive. If an individual polluted water, he was expected to purify it. Perhaps a look back can help us gain perspective on our use of water.

EARTH SCENES/E.R. DEGGINGER

△ **In the fall of 1991, eight women and men shut the door on the world and entered Biosphere 2, a sealed** terrarium of glass and steel about the size of two and one-half football fields. Biosphere 2 is intended to be self-sufficient with its members growing their own food, recycling air, water, and waste, and receiving only sunlight, electricity, and communications from the outside. The purpose? To serve as a model for pioneering other planets.

◁ **W**ater—heated by molten rock—will continue to provide power in areas where geothermal energy production is feasible. About 20 countries tap this resource. This geothermal plant is located in New Zealand.

43

△ **L**arge amounts of freshwater are locked up in polar ice. Towing icebergs from Antarctica to dry coastal areas such as southern California has been proposed. Outlandish! Impossible? Maybe. But exotic and imaginative solutions as well as traditional approaches will help solve tomorrow's water uncertainties.

Imagine turning over in bed at night and coming face to face with a shark ...or a barracuda ... a nightmare? No, the realization of a dream. The Jules' Undersea Lodge is named for Jules Verne, the author of "Twenty Thousand Leagues Under the Sea." This underwater hotel is 30 feet down in a protected lagoon off Key Largo, Florida. ▽

DICK DIETRICH

△ **Our water legacy should provide not only for clean and abundant water, but also for the preservation of water wilderness. Parks and preserves should continue to be set aside so that tomorrow's children might see their reflection in a crystal-clear mountain lake.**

A Water Legacy

How can an individual save clean water for the future? To get you started in your savings plan, here are a few simple ideas:

- Organize a stream clean-up project.
- Check your home and business for water leaks. Turn off all faucets and monitor your water meter for 30 minutes. If a dial moves, you have a leak.
- Test the water in your home or school to determine its quality.
- Organize a "community water conservation convention" to create awareness and knowledge of water-saving methods.
- Consider Xeriscape landscaping. That is, replace your thirsty lawn with drought-tolerant plants, shrubs, and ground cover.
- Turn off the water when you're not using it or else your savings will be doing down the drain!

SUGGESTED READING

BISWAS, ASIT K. *History of Hydrology.* New York: American Elsevier Publishing Company, Inc., 1970.

CADUTO, MICHAEL J. *Pond and Brook.* New Jersey: Prentice-Hall, Inc., 1990.

FREDERICK, KENNETH D. AND ROGER A. SEDJO (EDITORS). *America's Renewable Resources.* Washington D.C.: Resources for the Future, 1991.

MILLER, G. TYLER, JR. *Resource Conservation and Management.* California: Wadsworth Publishing Company, 1990.

POLIS, GARY A. *The Ecology of Desert Communities.* Tucson: The University of Arizona Press, 1991.

POSTEL, SANDRA. *Last Oasis, Facing Water Scarcity.* New York: W.W. Norton and Company, 1992.

RENNICKE, JEFF. *River Days, Travels on Western Rivers.* Colorado: Fulcrum Inc., 1988.

WHITAKER, JOHN O., JR. *The Audubon Society Field Guide to North American Mammals.* New York: Chanticleer Press, Inc., 1991.

Water brings together all living and nonliving things in a complex web of life on the Blue Planet. Water is a voyager in an immense journey around the planet. Powered by the sun, water is evaporated from the oceans and land or transpired from plants. Invisible, water vapor rises into the atmosphere. It cools and falls as precipitation. It collects in rivers or lakes as surface water or seeps into soil and becomes groundwater. Finally, it returns to the sea.

Water is an agent of change. Freezing, thawing, coursing, water can create rugged canyons. Changes may require minutes or millenniums. Water helps to shape the community in which animals live. A species' "water address" might be a tide pool or a vernal pond. Water works for all of us in industry, power generation, mining, navigation, fisheries, farming and ranching, and in fulfilling municipal domestic needs.

By accepting responsibility for our place in the water cycle, we will secure a promising social and economic future for the generations that will follow us.

◁ **Brilliant gold and red fall** foliage is reflected in the waters of Moose Pond, Maine. This celebration of color is a final salute to the balmy days of summer and a sharp contrast to the winter days ahead.

△ **W**ater boiling over boulders or whispering in gentle currents, the music of the river soothes the restless and adventurous soul. Some 20,000 people annually float the Colorado River through the Grand Canyon. This party is running the legendary Lava Falls. Seasoned boatmen call it, "the most exciting twelve seconds of your life."

"...I stand on deck, supporting myself with a strap, fastened on either side to the gunwale...striking a wave, she leaps and bounds like a thing of life, and we have a wild, exhilarating ride...." —John Wesley Powell

Books in the Story Behind the Scenery series: Acadia, Alcatraz Island, Arches, Biscayne, Blue Ridge Parkway, Bryce Canyon, Canyon de Chelly, Canyonlands, Cape Cod, Capitol Reef, Channel Islands, Civil War Parks, Colonial, Crater Lake, Death Valley, Denali, Devils Tower, Dinosaur, Everglades, Fort Clatsop, Gettysburg, Glacier, Glen Canyon-Lake Powell, Grand Canyon, Grand Canyon-North Rim, Grand Teton, Great Basin, Great Smoky Mountains, Haleakala, Hawai'i Volcanoes, Independence, Lake Mead-Hoover Dam, Lassen Volcanic, Lincoln Parks, Mammoth Cave, Mesa Verde, Mormon Temple Square, Monument Valley, Mount Rainier, Mount Rushmore, Mount St. Helens, National Park Service, National Seashores, North Cascades, Olympic, Petrified Forest, Redwood, Rocky Mountain, Scotty's Castle, Sequoia & Kings Canyon, Shenandoah, Statue of Liberty, Theodore Roosevelt, Virgin Islands, Yellowstone, Yosemite, Zion.

NEW: in pictures—The Continuing Story: Arches & Canyonlands, Bryce Canyon, Death Valley, Everglades, Glacier, Glen Canyon-Lake Powell, Grand Canyon, Hawai'i Volcanoes, Mount Rainier, Mount St. Helens, Olympic, Petrified Forest, Sequoia & Kings Canyon, Yellowstone, Yosemite, Zion.

This *in pictures* series is available with Translation Packages

Published by KC Publications • Box 15630 • Las Vegas, NV 89114

The child within us recognizes the magic of ▷ mixing raindrops and sunlight to create a rainbow. *Photo by Dick Dietrich*

Back cover: The waters of Oak Creek at ▷ Sedona, Arizona, refresh the land and the human spirit. *Photo by Dick Dietrich*

Created, Designed and Published in the U.S.A. Printed by Dong-A Printing and Publishing, Seoul, Korea Color Separations by Kedia/Kwangyangsa Co., Ltd.